YOUR KNOWLEDGE HAS VALUE

- We will publish your bachelor's and master's thesis, essays and papers

- Your own eBook and book - sold worldwide in all relevant shops

- Earn money with each sale

Upload your text at www.GRIN.com and publish for free

Bibliographic information published by the German National Library:

The German National Library lists this publication in the National Bibliography; detailed bibliographic data are available on the Internet at http://dnb.dnb.de .

This book is copyright material and must not be copied, reproduced, transferred, distributed, leased, licensed or publicly performed or used in any way except as specifically permitted in writing by the publishers, as allowed under the terms and conditions under which it was purchased or as strictly permitted by applicable copyright law. Any unauthorized distribution or use of this text may be a direct infringement of the author s and publisher s rights and those responsible may be liable in law accordingly.

Imprint:

Copyright © 2008 GRIN Verlag, Open Publishing GmbH
Print and binding: Books on Demand GmbH, Norderstedt Germany
ISBN: 978-3-668-04407-4

This book at GRIN:

http://www.grin.com/en/e-book/301441/factors-affecting-the-adoption-of-small-scale-irrigation-in-the-ameya-district

Tigistu Amsalu Oljira

Factors Affecting the Adoption of Small Scale Irrigation in the Ameya District of South West Shoa

GRIN Publishing

GRIN - Your knowledge has value

Since its foundation in 1998, GRIN has specialized in publishing academic texts by students, college teachers and other academics as e-book and printed book. The website www.grin.com is an ideal platform for presenting term papers, final papers, scientific essays, dissertations and specialist books.

Visit us on the internet:

http://www.grin.com/

http://www.facebook.com/grincom

http://www.twitter.com/grin_com

A RESEARCH REPORT ON
FACTORS AFFECTING ADOPTION OF SMALL
SCALE IRRIGATION IN AMEYA DISTRICT OF SOUTH WEST SHOA
ZONE-OROMIYA REGIONAL STATE
(The case of *Kulit* Small Scale Irrigation Project)

SUBMITTED TO
DEPARTMENT OF RURAL DEVELOPMENT
AND AGRICULTURAL EXTENSION
COLLEGE OF AGRICULTURE, HARAMAYA UNIVERSITY

IN PARTIAL FULFILMENT OF THE REQUIREMENTS FOR THE
DEGREE OF BACHELOR OF SCIENCE IN RURAL DEVELOPMENT &
AGRICULTURAL EXTENSIOIN

BY:-TIGISTU AMSALU OLJIRA

JULY, 2008

HARAMAYA UNIVERSITY

ETHIOPIA

Table of Contents

Acknowledgements

I would like to thank my advisor Ato Samson Ishetu for providing invaluable advice, comment & suggestions in writing my proposal and this report.

Acknowledgements go also to the South West Shoa Zone Agriculture & Rural Development Office, Ameya District Irrigation Development Office and Karayu Marii Development center for their support during conducting the survey.

Last but not least, I offer my special thanks to Mesi & Hawi (My wife & daughter) for their encouragement during my study and sacrificing their interest to let me join University.

List of Tables and Figures

List of Annexes

Acronyms (Abbreviations)

- CSA   Central statistics authority
- DA   Development Agent
- ESRDF   Ethiopian Social Rehabilitation and development Fund
- FAO   United nations Food and Agriculture Organizations
- Hec.   Hectare(s)
- MOI   Ministry of Information
- MOWR   Ministry of water Resource
- WUAS   Water users association
- OBPED   Oromia Bureau of Planning & Economic Development
- OIDA   Oromia Irrigation Development Authority
- O & M   Operation & Management
- PA   Peasant association
- SSIP   Small Scale Irrigation Project

1 Introduction

1.1 Background

Agriculture is the major primary economic activities of the Ethiopian population. Due to the existence of diversified topography, soil, weather and climatic conditions that favor agricultural activities the majority of the Ethiopian population have been engaged in and generate their income from the sector.

However, agriculture in Ethiopia is mainly characterized by the use of backward & traditional farm implements and subsistence farming system dominates by rain fed agricultural production that resulted to low and declining productivity of the sector.

Despite the bottlenecks mentioned above and a number of other factors, agriculture in Ethiopia was and remain to be the leading and determinant sector of the country's economy by being the major supplier of food crops for domestic consumption and the main sources of export commodities and capital to be accumulated for the process of establishing the future industrialized Ethiopia.

In order to make the sector plays its role to the level best, quite a number of efforts should be made by the concerned stakeholders to improve and change the overall performance of agricultural activities carried out on private agricultural holdings and large and medium scale farms (CSA, 2004).

In this direction, the government of the Federal Democratic Republic of Ethiopia has developed and started to implement a Rural Development policies & strategies; which focus on Agriculture and Rural Development. Then irrigation development and improving water resource usage is employed as one strategy to increase productivity (MOI, 2001). Though Ethiopia has 3.5 million hectares of irrigable land, irrigation covers only 0.16 million ha. or about 5% of the total irrigable land (MOWR, 1997). The dependence of most of the farmers on rain fed agriculture has made the country's agricultural economy extremely fragile and vulnerable to the impact of weather and climatic variability. Absence of rainfall during rainy seasons, and the consecutive dry spells at critical times of crop growing season lead to partial or total crop failure; which in turn results in food shortages.

Even though agriculture is one way or another has been the principal sector of the economy for several thousand years, it remained less efficient owing to old and traditional means and practices of farming on one hand, and the variability in the amount and distribution of rainfall from year to-year, on the other. Nearly, the entire peasant farming depends on rainfall which has greater impact in hindering the agricultural production system (Fekadu, 1991)

Thus, the use of supplementary irrigation from either traditional or modern water harvesting structures is considered the primary measure to be taken against the problem. In this direction government of Ethiopia is making serious efforts by allocating a fairly large amount of budget for the development of irrigation structures.

In the same way the **Kulit SSIP** have constructed by OIDA with the fund from ESRDF by diverting a river Kulit at the cost of ***1.8 mil. birr*** (*of this birr 122,221.70 was Community's contribution)* in the year 1992-1994 E.C. intended to serve about ***234 households*** and ***total population of 1170*** with objective of improving production.

In general to meet an objective of a given project the intended beneficiaries have to participate and make use of it, but there is case happened (as the one I have been studied) that beneficiaries would have not fully adopt or participate in a given technology.

The cause of this problem could be as explained on (water Report No. 17, FAO 1998) "a project – led approach in design and implementation is followed rather than a demand – drive selection. As a result, beneficiary participation is often sacrificed in favor of accelerating construction activities. Farmers are, therefore, not convinced of the need to take full responsibility for bringing land in to production & taking responsibility for scheme operation and Management." Thus, this particular study was intended to investigate factors hindering the farmer's participations on small scale irrigation while they have full access to it.

1.2   Statement of the Problem

As the main economic sectors of Ethiopia, agriculture is dependent on rainfall; this problem coupled with the lack of improved inputs and technologies leads the farmers to survive under pressure of food shortage.

Now a days, to over come or alleviate this problem different interventions has been taken of these promoting small scale irrigation development to use the country's surface and ground water was taken as a strategy and constructions are undergone with the participation of users.

In many cases, however there is some efficiency problems and (or management problem) is observed while farmers are willing to use irrigation. And those irrigation users have a good economic status.

But in this case large number of farmers under a command area of kulit small scale irrigation project lacks interest to use the technology, even though different measures have been taken to motivate them to participate. As the data from Ameya District reveals, out of the intended hectare to be developed under the irrigation, i.e. 200 hectares, only 34 hectares was developed as to year 1999 E.C

In the study area the cause for farmers' lack of interest in irrigation practice, and the factors affecting the adoption of this technology had not clearly identified. Therefore, this study was initiated to identify major problems which hinders farmer participation and to see factors affecting the adoption of small scale irrigation.

1.3   Objectives of the study

This research has conducted to attain the following objectives:-
1. to assess factors that hinders the adoption of Small Scale Irrigation technology,
2. to identify (examine) problems which contributed for less participation of the farmers on small scale irrigation (SSI).
3. to generate possible solutions to improve participation of the community on irrigation

1.4   Research Questions

The research was intended to answer the following questions:-
1. What are the factors impeding adoption of small scale irrigation?
2. What are the problems contributing for less participation of farmers on small scale irrigation?
3. What are possible solutions to improve community participation on irrigation?

1.5   Significance of the Study

As irrigation is the focus area of government's development intervention, different measures have been taken so far to mobilize farmers to participate in irrigation (the kulit small scale irrigation scheme); only few number of farmer are participate and the project is not functioning with its full capacity the resource is going lost. So to alleviate this problem the research has been conducted focusing:-

> On identifying major problems or cases that hinder farmers from using the irrigation,

> On evaluating factors affecting adoption of the irrigation development in the area,

> To recommend possible solutions which help to utilize this resources efficiently,

> To point out problems related to project planning, implementation and operation stages,

And, the study will contribute as a ground for further studies & has helped the researcher to get a practical experience on conducting a research. The research have generate scientifically analyzed information which will give a clue to improve farmers' participation on kulit SSIP-which in turn might have a great impact on productivity of the farm households.

2   Review of Literature

2.1   Definition of Concepts

An innovation is an idea, method or object which is regarded as new by an individual, but which is not always the result of recent research. (Vanden Ban & Hawkins,1996). And making use of a given technology needs some mental process to decide its adoption. As Roger (1962) defined adoption as the mental process through which an individual passes from first hearing about an innovation to final adoption. With regard to this idea, adoption process includes five stages:-

1. Awareness: First hear about the innovation,

2. Interest: Seek further information about it,

3. Evaluation: Weigh up the advantages & disadvantages of using it,

4. Trial: test the innovation on a small scale for yourself.

5. Adoption: apply the innovation on a large scale in preference to old methods. (Vanden Ban *et al*, 1996)

Roger also defines adoption as **a decision to continue use of an** innovation. This definition implies that the adopter is satisfied with the innovation. Roger Shoemaker (1971) defines adoption as a decision to make full use of new ideas as the best course of action available. As they indicate the adoption or rejection of an innovation is a decision by an individual. If he adopts, he begins using a new idea, practice, or object and cease using the idea he was using before the innovation. According to Dasgupta (1989), the term adoption refers to the continued use by individuals or groups of a recommended idea or practice over a reasonably long period.

In general there are different factors can be terraced that have negative impact on adoption of a technology. As Gautam (2000) describes it "The degree of non-adoption reflects, in part, the quantity and relevance of extension advice, especially given the technical, marketing, and resource constraints that farmers face."

2.2 Small Scale Irrigation

As the rainfall dependent agriculture is always risk prone and the average land holding decrement with the rapid population growth or pressure coupled with natural resource (soil) degradation; calls attention of policy makers and development practitioners to see for solutions to meet the production need through intensification of agriculture using improved technologies and promoting irrigation at small scale level to reach large number of people as the water resource of the area permits

FAO- defined small scale Irrigation development as the process of introducing effective water control techniques to schemes with an independent water supply and a command area not exceeding 500 hectares which are to be planned, developed and managed by farmers through the establishment of viable Water Users Association (WUAS) linked to existing local social structure. (FAO1998)

Also the same paper, (FAO,1998) describes that, "... to ensure farmer participation and to promote a feeling of ownership, which will lead to a commitment to undertake operation and management (O&M), emphasis is placed on substantial inputs and contribution from farmers towards the construction of the irrigation works".

If not implemented this way farmers are not convinced of the needs to take full responsibility for scheme operation and management.

In implementing a community managed small scale irrigation the Water User's Association (WUAS) have a great role by organizing the farmers to operate and manage the irrigation scheme which leads to better profitability. WUAS, have a direct impact on the performance of irrigation systems, along with technical, economic and government forces, but external factors also affects the structure and functioning of WUAS. So that it is better to examine these factors for correction & to ensure the sustainability of the project scheme (Madhur, 2000).

3 Methodology

3.1 Background of the study area

Ameya district is located in the South West Shewa Zone of Oromia Regional state. It has an area of 906.9km2, bordered by Tikur Inchini, Ambo, Wenchi, Weliso, and Nonno districts and the Regional State of Nations, Nationalities and Peoples of Southern Ethiopia. The district's capital *'Gindo'* is the only urban center & it is found on 30km from zonal capital Weliso & 144 km from Addis Ababa.

Physiographic ally, it is characterized by varied relief features, i.e. mountains, plateaus, hills, plains & valleys. The altitude of Ameya extends between 1600 and 3245 m.a.s.l, Mt. Roge (3245m) is the only known mountain in the district. The district is drained by Walga, Rome, Kulit Kela, Amigna Guda, Amigna Tika, Kulit Guda, Kabana and other 7 rivers as well as several seasonal streams. Ameya is classified into dega (40%) & weina dega (60%) agro climatic zones. Major soil types found in the district are chromic & octhic Luvisols (64% Chromic & Pellic vertisols 35.7%).

Woodland, Shrub & Bush, Savanna & man-made forests are found in the district. In the absence of protected forests & game reserves:- Baboon, Vervet Monkey, Duiker, Bush buck, Warthog, Leopard, Spotted hyena, Jakal, Pig & Civet cat are found dispersedly in the district.

As projected from the 1994 Population and Housing census the district had about 110243 population, of which 105832 were rural (53128 female) and 4409 population were urban (2469 female) population. Young, economically active and old age population accounted for 44.8, 51.2 and 4.0% respectively. The crude population density of the district is estimated as101 persons/km^2.

About 69.6, 10.9, and 5.5% of the district's total area were arable, grazing and forest lands respectively, while remaining was attributed to degraded, built-up and other areas. Average farmland and farm oxen holding sizes per farmer household were 0.3 ha. and 1.7oxen respectively. About 24.3% of the farmers were without single farm Ox. Teff, Barely, Maize, Wheat, Horse Bean, Vetch, Neug, Enset and Chickpea are the most widely cultivated crops in Ameya. The most commonly prevalent crop pests are aphids, army worm, stalk borer, etc. As the mixed farming is practiced in the area, the district has about 96,446 cattle, 16928 Sheep, 21344 Goats, 3121 Horses, 5541 Donkeys, 1240 Mule, 112170 Poultry & 10,317 Beehives /245 modern/.(OBPED,2000)

The specific study area, the *Kulit* small scale irrigation project is found in *Karayu Mari Sakala* PA, on the average altitude of 1860 m.a.s.l. about 7km far from Gindo town. The livelihood of the Community in the project/Research area is mainly dependent on mixed farming system in which Crop production has a lion share of farmers' source of income while the livestock production has a little but crucial contribution(as a traction power etc.).

With total area of 2700 hectares the current land use pattern of the project area (PA) is:-

Table 1:- Land use pattern of the Karayu Mari Sakala PA

Description	Area (ha)	%
Cultivated land	1944	72
Forest Land	151.2	5.6
Grazing Land	270	10
Village	178.2	6.6
Others	155.25	5.75
Total	*2700*	*100*

Source: Agricultural Development Center of the PA

According to the data obtained from PA's agriculture development center the population of the PA is:-

Table 2:-Population of the Karayu Mari Sakala PA

Description	Male	Female	Total
Households	568	17	585
Total population	2325	2012	4332
Family Size			7.4
Population density			$160.4/km^2$

Source: Agricultural Development Center of the PA

Table 3:- Major Crops produced & purpose of production in the Project area

Rain fed Crops	Purpose of production	Irrigated Crops	Purpose of production
Teff	*Consumption &market*	Onion	*Market &consumption*
Maize	*Consumption*	Tomato	*Market &consumption*
wheat	*Market*	Potato	*Market &consumption*
Nugi	*Market*	Banana	*Market*
Flax	*Market*	Coffee	*Consumption*
Vetch	*Market*	Chat	*Market*
Check pea	*Market &consumption*	Papaya	*Market*
		Avocado	*Market*
		Mango	*Market*
		Sugarcane	*Market*

Source: Agricultural Development Center of the PA

The Farming system of the area is a mixed farming which inclined towards crop production using rain water. But those farmers who have access to irrigation have an advantage of cultivating Vegetables & Fruits during the dry season of the year-which creates an opportunity to maximize their income

However Kulit SSIP; which were intended to irrigate 200 hectares & support about 234 HHs is the only irrigation scheme in this PA, the area cultivated since its construction have sowing a decreasing trend (*see Table*).

Table 4:- Trend of area Cultivated using Kulit SSIP

Year	Area cultivated (He.)	No. of beneficiaries	Production in Quintals	Remarks
1996	102	152	8160	
1997	58	205	2970	
1998	49.6	168	2825	
1999	34	145	2957	

Source:-Irrigation Development Office of Amaya District

3.2   Research Design

- Non experimental design

3.3   Sampling techniques

Stratified random sampling techniques were employed (to get information from both participating and non participating farmers)' and a total of *28* households *(11 users & 17 non-users)* were selected and interviewed this was about **12%** of intended beneficiaries or farmers accessed the irrigation facilities in the area. (*i.e 234 households)*

3.4   Methods of Data collection & source of relevant data

Relevant data have been collected from both primary and secondary sources. Such data as socio-economic condition, awareness and attitudes towards small scale irrigation, institutional support and wealth status of the respondent were covered. And both qualitative & quantitative data have been collected (from both primary & secondary sources).

a) Primary Data

The primary data on farmers' attitude & awareness towards the small scale irrigation scheme, and major constraints (external as well as internal) which hinders them to participate & their socio-economic status were examined from the sample households through scheduled interview in which both closed & open ended questions have been used.

b) Secondary Data

The secondary data regarding the case were reviewed from districts' office of Agriculture and Rural Development, Irrigation Development Office, Development Agent of the respective PA and other governmental institutions, and from different documents like report, project document etc.

In order to obtain additional information focus group discussions have been conducted with stake holders-PA leaders, community leaders, water users' committee, DA, irrigation experts, district administration bodies etc. Also observation methods were used to understand the general feature of the study area (physical and socio-economic conditions).

3.5   Method of Data analysis

The initiative of the study was to examine factors affecting or hindering farmers to participate in the SSI in the area, the qualitative data have been analyzed by **comparison (compare & contrast), explanation, discussion, interpretation** etc. And for quantitative data the **descriptive statistics-** mean, frequency percentage etc. were employed.

4   Results and Discussion

4.1   Summary of Background Information of Sample Farmers

Table 5 Average family size, Age & Land holding of sample farmers (N=28)

No	Description	Average	Minimum	Maximum
1	Family size	6.5	3	13
2	Land holding /households in hec.	2.02	0.125	4

Source: - Own survey (2000)

According to the survey result revealed in the Table 1, the average family size of the respondents is 6.5, while the average land holding is only 2.02 hectares per household, which is less than the average of the districts' holding size.

Table 6 Educational status of Respondents (the sample farmers)

No	Educational level	Users (N=11) Frequency	%	Non-user (N=17) Frequency	%	Total No. of respondent Frequency	%
1	Illiterate	3	20	5	33.3	8	28.5
2	1-4(basic education)	3	30	6	35.3	9	32
3	5-8	3	30	5	33.3	8	28.5
4	9-10	2	20	1	5.8	3	10.7
5	11-12	0	-	0	-	0	-
6	>12	0	-	0	-	0	-
	Total	11		17		28	

Source:- Own Survey(2000)

Among sample users only **20%** are illiterate but of the total non users **33.3%** comprises illiterate.

Table 7 Frequency distribution of age class of sample farmers

S no	Type of respondent (Description)	Age class of sample farmers			Remarks
		18-35	36-49	Above 50	
1	Users(N=11)	5	3	3	
	%	45.5	27	27	
2	Non users(N=17)	5	7	5	
	%	29.5	41	29.5	

Source:- Own Survey(2000)

4.2   Access to media & Source of Agricultural information of the sample farmers

Table 8 Frequency of mass media Possession (access) of sample farmers (N = 28)

Type of media	Users (N=11)		Non users (N=17)		Total (N=28)		Remarks
	Frequency	%	Frequency	%	Frequency	%	
Radio	9	81%	9	52.9%	18	64.28	
TV		0	0		0		
Telephone		0	0		0		

Source:- Own Survey (2000)

According to the survey *81%* of the sample farmers who are users have access to Radio (mass media) while only *52.9%* of those non-users have access to mass media.

Table 9- Agricultural Information sources of sample farmers

No	Sources of information	Users (N=11)		Non-users (N=17)		Remarks
		Frequency	%	Frequency	%	
1	Mass media (Radio)	3	27%	8	47%	
2	Development Agents	11	100%	17	100	
3	Contact farmers	0	-	6	35.3	
4	Friends (Relatives)	1	9%	8	47	

Source:- Own survey (2000)

All users & Non-users of the SSIP are obtaining agricultural information from development agents. Additionally 27% of users & 47% non users obtained information from mass media (Radio).However, *81%* of non-users have access to radio, and only *27%* of them have obtained agricultural information from it while *47%* of non users have access to radio.

4.3 Impact of Irrigation on living status of sample farmers

Table 10– Comparison of Oxen owned by the sample farmers

S no	Description	No. of Oxen owned by the sample farmers					
		0	**1**	**2**	**3**	**4&above**	**total**
1	Users(N=11)	2	1	5	1	2	11
	%	*18*	*9*	*45.5*	*9*	*18*	*100*
2	Non Users(N=17)	4	1	10	1	1	17
	%	*23*	*5.8*	*58.8*	*5.8*	*5.8*	*100*

Source:-Own survey (2000)

This table revels that among 11 sample users of the Kulit SSIP only 18% are with out a single Ox, while the rest 82% owned an Ox (of which 9%,45%,9%,&18%)of farmers possess 1,2,3,& 4 Oxen respectively). But 23% of those farmers who are non users have no a single Ox, and those who have more than 2 Oxen are about 11%. This means that, those users of Irrigation are economically better off.

Table 11 - Response of users regarding the economic impact of the project (N=11)

No	Income increasing by	Frequency	%	Remarks
1	Two fold (Doubled)	2	18	100% increase
2	One third	7	64	33.3% increase
3	One fourth	2	18	25% increase
4	Not increased	0	-	
5	Not Responding	0	-	
	Total	11	100	

Source:- Own survey (2000)

According to the response of sample farmers the annual income of all the users of the irrigation project has been increased since they are using it. And this indicates roughly the positive return obtained by users for using the scheme

Table 12- Attitudes of sample farmers towards the impacts of Irrigation project (N = 28)

No	Advantages matrix	Frequency	%	Remark
1	Helps to generate additional income for the family	25	89.3	
2	Improves Nutritional status of the family	25	89.3	
3	Helps to produce two times a year (increases production)	25	89.3	
4	Has negative impact on the environment (*seepage affecting a farm land*)	3	10.7	
5	Creates conflict with the community around the head work site, which are not users	1	3.6	
6	Not responding	2	7	

Source- Own Survey (2000)

As it is reveled in the above table, **89.3%** of all the sample farmers have a positive attitude towards the SSIP, even those who at the time of survey not using the scheme gave a positive response regarding its impact. This is because the project (scheme) enables the users to maximize their income through producing tow times a year & in turn improves nutritional status of the family. And also **14.3%** of the sample farmers have a negative attitude towards the kulit SSIP. Of these 10.7% of sample farmers indicate the negative impacts due to seepage problem which affects their farm land and also 3.6% of them point out the conflict with community around head work site as a negative impact.

4.4 Attitude of sample farmers towards development of the scheme & their condition on using irrigation

Table 13-Status of sample farmers on using Kulit SSIP (N= 28)

Status of sample farmers	Frequency	%	Remarks
Using now (N= 11)	11	39.3	
Non users (N=17)	17	60.7	
Discontinued using	14	82.35	Out of the total non-users
* Never used	3	17.65	"

Source:- Own survey (2000)

Out of the total sample farmers only 39.3% are using Kulit SSIP and the balance are non-user. About 82% of non-users are those who have been users of the scheme but now terminated (discontinued) using, while the rest 18% were never used.

Table 14-Interest of the sample farmer for Development (construction) of Kulit SSIP

Sample farmers	Had interest		Had no interest		Had contribution	
	frequency	%	frequency	%	frequency	%
Users (N=11)	11	100	0	-	11	100
Non-users (N=17)	17	100	0	-	17	100
Total	**28**	**100**	**0**	**-**	**28**	**100**

Source:- Own survey (2000)

All respondents have had an interest for the development of the Kulit SSIP. And also all the sample farmers have participated during construction by contributing 10% community share in cash and labor. The reasons for the interest according to the sample farmers are:-

- As the area is venerable to livestock endemics, irrigation will help to balance the loss,

- It will help to double crop intensity (to cultivate vegetables two times a year),

- Since the climate of the area (rain fall) is changing through time, it is vital to develop irrigation schemes to increase production & productivity.

- To get additional income.

ᮬ They have ample evidences around: that, the livelihood of those farmers using irrigation have been changed or improved.

Table 15- Access to Farm land under command area (N=28)

No	Holding size (ha)	Users (N=17)		Non-users (N=17)		Remarks
		Frequency	%	Frequency	%	
1	0			1	5.8	
2	<0.25	1	9	0	-	
3	0.25-0.5	6	54.5	11	64.7	
4	0.6-0.75	0	-	0	-	
5	0.76-1	3	27.3	3	17.6	
6	1.01-1.25	0	-	0	-	
7	1.26-1.5	1	9	1	5.8	
8	1.6-2	0	-	0	-	
9	2.01 & above	0	-	1	5.8	
	Total	**11**	**100**	**16**	**100**	

Source:- Own survey (2000)

As it is shown in the above table all the sample farmers have a farm land under command area but only one HH *(5.8%)* of sample farmers from Non users have no Land under command while *94%* of non users have a farm land in the command of Kulit SSI scheme.

The average holding is 0.69 hectares per HHs- with the minimum & maximum holding of 0.125 & 3 hectares respectively.

Table 16- Participation on Operation & maintenance of the project (Users N= 11 & Non users N= 17)

No	Mode of participation	Users		Non users	
		frequency	%	frequency	%
1	Labor (clearing canals, maintaining structures)	11	100	0	-
2	Contributing money (a yearly fee for different expenses)	6	54	0	-

4.5   Major constraints that were hindering the sustainable use of irrigation Scheme

Table 17- Factors that hinders usage of Kulit SSIP (Reasons given by non-users for discontinuing using) N=17

No	Possible Reasons for not using	Frequency (N= 17)	%	Remarks
1	Has less return			
2	Shortage of irrigation water (due to seepage, distribution...)	16	94	
3	Structural problems,	9	53-	
5	Lack of Agricultural inputs	7	41	
6	Less financial capacity of farmers	0	-	
7	Inequitable water management & distribution	13	76.5	
8	Shortage of farm land under command area	1	5.9	
9	Lack of knowledge & information	0	-	
10	Market problem	0	-	
11	Distance from the Scheme	2	11.8	
12	Have access to other irrigation schemes	0	-	
13	Pest & disease Problem	0	-	
14	Shortage of labor force	2	11.8	
16	Location of farm land	2	11.8	
17	Salinity problem	0	-	

Source:-own survey(2000)

As it is described in the above table; the major reasons stated by the interviewed non-user sample farmers were:-

- Shortage of irrigation water (b/c of seepage in earthen canals, breakage of lined canals,& application problems)

- Problem of water management & distribution (Unfair distribution of water among users, bylaws are violated, corruption)

- Distance from the site

- Shortage of Labor force to work on irrigation,

- Lack of Agricultural inputs, since there is no provision of inputs for the program (irrigation);

are the same with reasons stated by the sample farmers during focus group discussion .

Table 18:-Area cultivated by Sample farmers as to year 2000

Sample farmers	Total farm land owned in the command (hec.)	Cultivated as to year 2000(hec.)	%age
Users	7.125	2.887	40.5
Non-users	12.25	0	-
Total	**19.375**	**2.887**	**14.9**

Source:- Own survey (2000)

Out of the total Farm land owned by the sample user farmers in the command only 40.5% was cultivated as to year 2000. But if the total farm land owned by all sample farmers were considered the area cultivated will be 14.95% only.

Table 19- Frequency distribution of Area cultivated by users as to year 2000 (N=11)

S.No.	Class of Land Developed	Frequency	percentage
1	0	2	18.18
2	<0.25	3	27.3
3	0.25 – 0.5	5	45.5
4	0.6 – 0.75	1	9
5	0.76 – 1.0	0	-
6	1.01 – 1.25	0	-
7	1.26 - 1.5	0	-
8	1.6 - 2.0	0	-
9	2.01 & above	0	-
	Total	11	100

Source:- Own survey (2000)

About 45.5% of those sample users were cultivated < 0.25 hectare of land, which is less than the recommended rate of farm land to be developed per HHs.(i.e. 0.5 hectare according to OIDA)

Fig.1 Area Cultivated by users sample farmers (1997-2000)

Source:- Own survey (2000)

According to the survey even the area cultivated by users for the last 4yrs shows decreasing trend. In order to understand the causes of this decreasing trend those sample farmers who cultivates a land < 0.5 hectares in this year were asked to reason out and the following facts were revealed. (see the table below).

Table 20- Response of users about factors which leads them to decrease cultivated Land

No	Possible Reasons for decreasing cultivated land	Frequency (N=11)	%	Remarks
1	Lack of Agrl. Inputs	4	36.4%	
2	Market problems	0	-	
3	Inadequate marketing infrastructures (storage)	0	-	
4	Shortage of Labor	1	9%	
5	Financial problem	3	27.3%	
6	Lack of knowledge & information (on irrigation)	1	9%	

7	Shortage of farm land	3	27.3%	
8	Shortage of Irrigation water	6	54.5%	
9	Problem of water mgt. & distribution	6	54.5%	
10	Pest & Disease problem	1	9%	

Source: Own survey (2000

According to the survey, area cultivated by those user sample farmers have been decreased due to the following reasons:-

- shortage of irrigation water as a result of structural & management problems,

- inequitable water distribution because of the managerial problem the WUA had,

- shortage of inputs required as there is no supply of inputs for the program(irrigation),

- financial problems since they have no access to credit facilities in their area,

- Lack of access to irrigable land in the command ,&

- Pest & disease problems etc.

5   Conclusion & Recommendations

5.1   Conclusion

- ❖ Shortage of irrigation water due to structural problems (seepage in canals) & management problems are the major constraints for participation.
- ❖ However those users are economically better off, they were not cultivated a proper Land size sustainably / continuously due to different reasons (social, technical, economic, etc.)
- ❖ Young age groups of sample farmers are relatively active participant (adopter) of irrigation project.
- ❖ Those users have more access to mass media & they   were less illiterate in comparison with non users.
- ❖ All the farmers in the area (whether they are using or not using now) have had an interest & participation during the construction of the scheme.

* In addition to those drop outs (terminated using); even the area cultivated by users has showed a decreasing trend for the last four years. -Area cultivated by users of the scheme was less than the recommended rate of Land to be managed per households.(i.e. 0.5 he.)

* About 89% of all the sample farmers have a positive attitude towards the project, and also those users have wittiness the positive economic return they enjoyed by using the scheme.

* Irrigation has advantage if properly managed &used:-
    - by increasing production &productivity per unit area,
    - by generating additional income to the family & improving nutritional status.

* Additional incomes obtained from the project help the users to save or store their crops for hard times, & also help to buy farm animals (oxen etc.) or build asset

* Lack of well organized water users' association was a major cause for inequitable water distribution- which in turn decreases the adoption & participation.

* The technical/structural problems occurred (lack of maintenance) was the main limiting factor on using the scheme.

* Interest & participation demonstrated during a project construction; will not be sustained unless other components of the technology like inputs, credit facilities etc. were provided throughout implementation period.

* Contribution from farmers during construction of irrigation works, which assumed to develop a sence of ownership will not be a guarantee for sustainable undertaking of operation & management of the scheme, unless continues correction measures taken by respective bodies while d/t problems occurred throughout implementation period.

* However the Land holding size of households in the area is very small ,which should push them to use irrigation in order to increase productivity per unit area, many of the farmers in the command area terminated using with scheme- thus we can conclude that technologies which needs a collective decision are very difficult to maintain a sustainable adoption.

5.2 Recommendations

☞ All necessary agricultural inputs must be available at the time of planting with irrigation. (Arrange credit facilities for irrigation).

☞ There must be clear demarcation between the users & government regarding the maintenance of the scheme (Who is responsible for what type of maintenance?).And those structural problems like seepage in unlined canals, and broken canals should be addressed urgently.

☞ Practical training on irrigation should be given to technical staff of extension, development agents & the farmers themselves.

☞ Re-organizing the water users association with the full participation of all intended beneficiaries of the project, & develop their managerial ability through continuous training.

☞ Empowering the water users association till they can stand by their own is very crucial to achieve the intended benefits from a given irrigation scheme

☞ And it is better to develop the WUAs in to coops to have legal personality which helps to develop their bargaining power in the market.

☞ Those farmers who have more than 0.5 hec. of land under command; which is difficult to manage per household must change the surplus for those farmers having no land in the command to irrigate. (According to the OIDAs' directives on irrigation Land use).

☞ Farmers need to be properly instructed in irrigation technology & encourages to regularly contribute money & communal labor for the operation & maintenance of the irrigation scheme.

☞ In order to irrigate more potentially irrigable land existing users should manage water more efficiently; thus promotion of water saving technologies based on site specific condition is vital (like mulching, lining of canals, drip irrigation , etc.)

☞ It is better to assign an expert of small scale irrigation to the site permanently to help the beneficiaries on managing the scheme & make proper use of it.

☞ In general, extension service delivery to beneficiary households needs to be improved& strengthened through: training in irrigation water applications, creation of demonstration plots, use of compost manure, farmer exchange visits, and improved post-harvest technology.

References

A.W.Van den Ban & H.S.Hawkins, 1996.Agricultural Extension.2nd edition. Blackwell science Ltd, Germany

Council of the Regional state of Oromiya-Bureau of planning and Economic development, Socio-Economic Profile,2000, Addis Ababa, Ethiopia.

CSA,2004,Ethiopia Agriculture Census 2001/02 for large and medium scale farms. Addis Ababa, Ethiopia.

Dasgupt.S.1989,Diffusion of Agricultural innovations in village India. Department of Sociology& Anthropology. University of prince Edward Island, Canada.

FAO,1998, Water Report No.17

Fekadu Tilahun,1991.Economics of Irrigation planning. Prospect for Resource use optimization; The case of Hidi Irrigation Project, Msc Thesis, Alemaya   University, School of graduate Studies.

Madhur Gautam, 2000.An impact Evaluation, World Bank.

Ministry of Information,2001, Federal Democratic Republic of Ethiopia; Rural development policy, strategy& directives(*Amaharic version*), Addis Ababa, Ethiopia.

Ministry of water Resources (MOWR), 1997 Water and Development quarterly bulletin, 1 (4): June 1997, Addis Ababa, Ethiopia.

Oromia Bureau of planning & Economic Development, March, 2000 – Addis Ababa, Ethiopia.

Rogers, E.M, 1962 Diffusion of Innovation. The Free press of Glencone. 367p.

Annex 1:-

Questionnaire (factors *affecting adoption of Small scale irrigation- the case of kulit SSIP)*

1. Personal Information

- Name of Respondent ____ Religion_________

                Age __________   **Sex**   Male ☐   Female ☐

- Number of Families

            Male _______________
            Female ____________
            Total ______________

- Land holding

            Farm land (Ha.)____________

            Grazing land(Ha.)__________

            Forest land (Ha.) ___________

            Irrigable land(Ha.) _________

            Total (Ha) ______________

- Livestock possession (Quantity)

            - Cattle ___________________
            - Pack Animal _____________
            - Poultry _________________
            - Beehives ________________

- Housing condition

            - Corrugated sheet of Iron _________
            - Tukuls (traditional) _____________

- Educational background of the Household heads

➡ Man/women -

            ☐ 1-4

            ☐ 5-8

            ☐ 9-10

            ☐ 11-12

            ☐ Others

Questionnaire (*factors affecting adoption of Small scale irrigation- the case of kulit SSIP*)

- What social assignment do you have in the society/or government

☐ Idir leader   ☐ Religion leader

☐ Administration

☐ Local Leader

2. Did you have an interest for the implementation of the *Kulit SSIP* ?
        ☐ Yes   ☐ No

2.1. If Yes/No Why?

____

____

____

3. Did you have a contribution/participation during the project planning & construction?

☐ Yes   ☐ No

3.1 If yes how/ if no why?

____

4. Do you have a farmland under the command area of the project?

☐ Yes   ☐ No

5. Did you participate in irrigation (of *Kulit SSIP*)

☐Yes   ☐No

5.1. If yes, how much Ha. of land did you have developed in the last 3 years:-

        1997_______________Ha.
        1998 _______________Ha
        1999 _______________Ha

5.2. Which type of crops did you mostly irrigated?

☐Vegetables

☐Fruits

☐Field crops

☐Others

5.3. Have the irrigation scheme contributed to an increase in your income?

☐ Yes   ☐ No

Questionnaire (*factors affecting adoption of Small scale irrigation- the case of kulit SSIP)*

5.3.1 If yes, how much do you think is the increase in your income?(in contrast without the project)

     ☐ Less than one fourth of the income

     ☐ One- third of the income

     ☐ Tow-third of the income

     ☐ More than double

5.4. If your answer for Question No.5 is NO, Why didn't you produce using the irrigation?

     ☐ Has less return (Productivity)

     ☐ Shortage of irrigation water

     ☐ Lack of agricultural inputs

     ☐ Financial capacity of the farmers

     ☐ Problem of water management & distribution

     ☐ Shortage of farmland under command area

     ☐ Lack of knowledge & information (on irrigation)

     ☐ Market problems

     ☐ Distance from the scheme

     ☐ Have accesses to other irrigation scheme

     ☐ Pest and Disease problem

     ☐ Problem of land re-allocation within command area (command was not delineated)

6. Have you participated on the maintenance & management of the project structures?

6.1 If yes, what was your contribution in maintenance & management of the project structures?

____

6.2 If No, why?

____

7. What do you think of the impact/effect of the project on your livelihoods?
   - Positive

____

   - Negative

____

Questionnaire (*factors affecting adoption of Small scale irrigation- the case of kulit SSIP*

8. Where did you get agricultural information?

☐ Mass media (radio, TV)   ☐ Contact farmers (neighbors)

☐ Development agents

9. Have you getting technical assistance from agricultural experts?

      Yes ☐   No ☐

9.1 If yes, how often did you meet these agricultural experts?

      ☐ Once a week

      ☐ Twice a week

      ☐ Once fortnights

      ☐ _____________

9.2 If no, why?

      ☐ I have no trust

      ☐ The experts are not willing to help as

      ☐ Other reason (if any)___________________

YOUR KNOWLEDGE HAS VALUE

- We will publish your bachelor's and
  master's thesis, essays and papers

- Your own eBook and book -
  sold worldwide in all relevant shops

- Earn money with each sale

Upload your text at www.GRIN.com
and publish for free